I0482875

The lifePhone

J.E. Angel, Jr.

Copyright © 2014 J.E. Angel, Jr.

All rights reserved.

ISBN-13: 978-1499548631

ISBN-10: 149954863X

WHAT IF ...

... the ENTIRE universe was a smart phone, and everything on it was an app, a program, or a computer code? ...

... and the Programmer of the smart phone allowed some free will to happen with the apps and data content inside the smart phone that set off all sorts of problems? ...

... and the Programmer actually step foot inside the smart phone (becoming part of the apps and data himself) alongside the other apps and data to fix everything? ...

... and the Programmer taking this radical step actually fixed the smart phone and made it perfect in the end?

CONTENTS

1
The Perfect Beginning
and The Ultimate Virus Attack

John 1:1-2

Way back when it all started, it was just The Powerful, All-Knowing Programmer and The Incorruptible Source Code. The Powerful, All-Knowing Programmer and The Incorruptible Source Code were in perfect synch with each other and were both there when it all started.

John 1:3

The Powerful, All-Knowing Programmer and The Incorruptible Source Code created the lifePhone. That's how everything got here—sky, grass, trees, flowers, animals, water, people, planets, stars. All of it was by His grand design and power. None of it was by accident.

Genesis 2:1-3

lifePhone a success! The lifeOS completely up and running! The sky, grass, trees, flowers, animals, water, people, planets, stars, etc. rock! The lifePhone was built in six rotations of the Earth's axis. The Powerful, All-Knowing Programmer and The Incorruptible Source Code decided to rest for the seventh rotation of the Earth's axis to set an example for the humans that He designed as part of the relationship that He wanted with them. The Powerful, All-Knowing Programmer and The Incorruptible Source Code defined these seven rotations of the Earth's axis as a week and wanted the humans to stop and focus their worship on Him on the seventh day.

Genesis 2:15-17

The Powerful, All-Knowing Programmer and The Incorruptible Source Code directed the first human that He created to make his pad in a fabulous garden called Eden. The Powerful, All-Knowing Programmer and The Incorruptible Source Code gave the human responsibilities there to look after the garden. The Powerful, All-Knowing Programmer and The Incorruptible Source Code also pointed out a tree in the garden. That tree was off limits. The human could eat from all the other trees which all had scrumptious flavors to choose from. If the human ate from the tree that was off limits, it would set off a chain reaction of death, destruction, and all that is bad. Basically, it would set off the mother of all computer viruses and mess up the lifeOS.

Genesis 2:18-25

The Powerful, All-Knowing Programmer and The Incorruptible Source Code did not want the human to be alone. After all, The Powerful, All-Knowing Programmer and The Incorruptible Source Code had created male and female of all the animals and told the first human to name them all. The Powerful, All-Knowing Programmer and

The Incorruptible Source Code wanted the man to have a helpful companion. The Powerful, All-Knowing Programmer and The Incorruptible Source Code caused the man to fall asleep and performed the first surgery. One less rib later, the man tripped out on the finest being in creation and got all excited because she came from him and was like him, but different in all the good ways. They lived in a nudist colony and didn't think badly of it.

Genesis 3:1-19

Things were going great until the Evil Accusing Deceiver and Corruptor took on the form of a snake and started throwing out all sorts of questions and deceptions.

Surely The Powerful, All-Knowing Programmer and The Incorruptible Source Code didn't mean they couldn't eat from that one tree?

The Powerful, All-Knowing Programmer and The Incorruptible Source Code must be trying to keep something good from them!

This has to be a conspiracy!

The Powerful, All-Knowing Programmer and The Incorruptible Source Code are trying to keep them from enjoying something that He made.

The humans thought about it. It did look like good fruit on that tree. The snake was trying to show them how they could get an upgrade by eating the forbidden fruit for it would make them wiser. They gave in and ate the forbidden fruit.

Suddenly, there was a chain reaction of death, destruction, and all that is bad! lifeOS was compromised! The virus spread like crazy causing all the perfection of the lifeOS to go haywire!

For starters, the humans began feeling bad about the whole nudist colony thing so they made some aprons out of fig leaves.

Then, the humans felt the need to hide when The Powerful, All-Knowing Programmer and The Incorruptible Source Code came to visit them like He was in the habit of doing. The Powerful, All-Knowing Programmer and The Incorruptible Source Code asked where they were. The man came clean and said they were hiding because the naked thing was a problem. The Powerful, All-Knowing Programmer and The Incorruptible Source Code asked if they had eaten the forbidden fruit. The humans admitted they had.

This led to a trial in which The Powerful, All-Knowing Programmer and The Incorruptible Source Code was the judge, jury, and executioner. Snake was cursed. Humans, especially women, would hate Snake and want him dead. Having babies would go from being a piece of cake to being a major pain. Plants and animals that had been pleasant to be around would become dangerous. The eloquent buffet was over! No more eating all the scrumptious fruit from the trees! The man would have to work to have food, and work would be unpleasant as all get out!

Genesis 4:1-16

The two humans enjoyed the special physical and emotional connection that The Powerful, All-Knowing Programmer and The Incorruptible Source Code designed to happen between a man and woman who are married. This resulted in the birth of their first son. Some time later, they gave birth to their second son. As the boys grew, the elder brother had a major green thumb going on while the younger brother had a knack for being a shepherd.

As time progressed, The Powerful, All-Knowing Programmer and The Incorruptible Source Code taught the humans an element of worshipping Him that involved sacrificing their best animals. The blood of these animals symbolized the perfect blood sacrifice that The Powerful, All-Knowing Programmer and The Incorruptible Source Code required to pay for all the damage caused by The Virus when they ate the Forbidden Fruit. The Powerful, All-Knowing Programmer and The Incorruptible Source Code would send this perfect sacrifice at the right time. Until then, The Powerful, All-Knowing Programmer and The Incorruptible Source Code wanted the humans to sacrifice their best animals to symbolize the perfect blood sacrifice that would one day take place.

As the time of worship began, the two brothers brought two different worship styles to the table. The elder brother brought his best vegetables (from his green thumb skills) which was not following directions given by The Powerful, All-Knowing Programmer and The Incorruptible Source Code. The younger brother followed directions and brought his best animals to sacrifice. The Powerful, All-Knowing Programmer and The Incorruptible Source Code accepted the younger brother's sacrifice but rejected the elder brother's sacrifice because he did not follow directions and was just going through the motions with worship. Enraged, the elder brother trapped the younger brother and murdered him in a fit of psychotic rage and jealousy. The Powerful, All-Knowing Programmer and The Incorruptible Source Code came looking for the younger brother. The elder brother lied about the younger brother's whereabouts and mumbled some trash talk about it not being his job to keep up with his brother. The Powerful, All-Knowing Programmer and The Incorruptible Source Code sentenced the elder brother to lose his green thumb skills, have to roam around the Earth as a restless wanderer, and bear a mark that he was not to be punished by death for being a murderer.

Genesis 6:1-7

The Earth went through a big population increase. Some evil angels who had tried to overtake The Powerful, All-Knowing Programmer and The Incorruptible Source Code but failed to defeat Him came to Earth and interacted sexually with the women on Earth. This caused demonized people to be all over the place which intensified The Virus. The Powerful, All-Knowing Programmer and The Incorruptible Source Code looked down and hated the mess that had spiraled so out of control. He couldn't let the demonized people completely take over the Earth and further mess up the lifeOS. He made a plan to destroy everything—well almost everything.

Genesis 6:11-14

The Powerful, All-Knowing Programmer and The

Incorruptible Source Code found a cat named Noah who tried to follow all the rules for worship and proper behavior. Noah, his wife, their three adult sons, and the ladies these sons married were all free from the contamination from the evil angels. They were 100% human and trying to obey The Powerful, All-Knowing Programmer and The Incorruptible Source Code. The Powerful, All-Knowing Programmer and The Incorruptible Source Code instructed Noah to build a huge boat that would go by the name, The Ark. It was gonna rain for the first time ever 'cause The Powerful, All-Knowing Programmer and The Incorruptible Source Code was gonna change the hydration cycle for the Earth.

Genesis 7:7-10

Noah and the peeps in his family followed the directions that The Powerful, All-Knowing Programmer and The Incorruptible Source Code gave them. Since The Powerful, All-Knowing Programmer and The Incorruptible Source Code made all the critters, the animals had to do what The Powerful, All-Knowing Programmer and The Incorruptible Source Code said to do. Interestingly, when The Powerful, All-Knowing Programmer and The Incorruptible Source Code gave the animals specific directions to follow, they always obeyed—unlike the humans. It began to rain for the first time ever plus water burst forth from within the Earth until only those inside The Ark survived.

Genesis 11:1-9

The Powerful, All-Knowing Programmer and The Incorruptible Source Code did a reboot of the lifeOS with Noah's family and the animals on The Ark. With the extermination of the evil angels and their demonized, part human offspring, the Earth was back to human beings, animals, and plants. Time went by until there were lots of people on the Earth again. The people all stayed together in one place and wanted nothing to do with the instructions that The Powerful, All-Knowing Programmer and The Incorruptible Source Code had given them. In fact, they decided to build a tower up to The Powerful, All-Knowing Programmer and The Incorruptible Source Code. Since they all spoke the same language, this proved to be easy for them. Not only were they building their way to The Powerful, All-Knowing Programmer and The Incorruptible Source Code, they wanted to create a room for Him at the top and demote Him from his powerful status as a "big H" Him to a "little h" him. In this room, they would give "him" a bed and take food to "him." With The Powerful, All-Knowing Programmer and The Incorruptible Source Code "needing" them, they could complete their hard-corps attitude of disobeying all instructions that The Powerful, All-Knowing Programmer and The Incorruptible Source Code had given them. The Powerful, All-Knowing Programmer and The Incorruptible Source Code refused to let this happen and threw The Language Disruption Algorithm at them. Unable to communicate clearly with one another, they abandoned the project. and spread out all over the Earth.

2
The Creation of The Lead App

Genesis 12:1-3

The Powerful, All-Knowing Programmer and The Incorruptible Source Code wanted to bless the human beings, but the human beings had to get their attitudes right about treating Him like a "big H" Him instead of a "little h" him. He found a cat named Abram who was living up the high life with lots of money from a prosperous family business located in a nice climate. The Powerful, All-Knowing Programmer and The Incorruptible Source Code told Abram to take his immediate family, leave his extended family, and go on a journey. This journey would result in Abram becoming a great nation. Becoming a great nation wasn't about Abram, though. Instead it was about The Powerful, All-Knowing Programmer and The Incorruptible Source Code getting the respect, good press, and props they deserved. If Abram would follow the instructions as he received them from The Powerful, All-Knowing Programmer and The Incorruptible Source Code, this Lead App would be a success.

Genesis 17:2-7

The Powerful, All-Knowing Programmer and The Incorruptible Source Code started making hard-corps promises to Abram. Abram's name became Abraham to commemorate all the promises. The first promise was having many descendants and always being their specific Powerful, All-Knowing Programmer and their specific Incorruptible Source Code.

Genesis 15:18-21, Genesis 17:8

The Powerful, All-Knowing Programmer and The Incorruptible Source Code promised Abraham that his people would have a specific plot of real estate on the Earth. This area came to be known as Israel. Abraham's people came to be known as the Jews. They have not yet acquired all the land promised to them, but The Powerful, All-Knowing Programmer and The Incorruptible Source Code will keep His promises and make this happen in His perfect timing. Again, the point of The Powerful, All-Knowing Programmer and The Incorruptible Source Code giving Abraham descendants and land was to define the Lead App (a special people to bless who would also bless the rest of the people on the planet). The land would be a headquarters for The Powerful, All-Knowing Programmer and The Incorruptible Source Code to carry out His operations.

Hebrews 11:8-16

Abraham tried to believe and obey what The Powerful, All-Knowing Programmer and The Incorruptible Source Code told him to do. Many times, it didn't make sense.

Yet he got his head in the game and went on the journey anyway. Living on the edge with such blind faith meant more to Abraham than having all the details and the securities of a fancy place to hang his hat. Abraham's son, Isaac, caught on to the whole Faith Factor thing as did his son, Jacob. They knew The Powerful, All-Knowing Programmer and The Incorruptible Source Code had a way cool cityscape waiting for them if they would just press on. None of this was easy.

Abraham and his wife were really old but had no children of their own. The Powerful, All-Knowing Programmer and The Incorruptible Source Code promised them a child which finally came in their old age. With such odds against them in their old age, The Powerful, All-Knowing Programmer and The Incorruptible Source Code did bless them with the child who led to grandchildren who led to great-grandchildren who led to more and more people until there were lots of peeps who could trace their ancestry back to Abraham and see that The Powerful, All-Knowing Programmer and The Incorruptible Source Code kept His promises!

Abraham and his peeps never saw the full extent of the land, city, and other details that The Powerful, All-Knowing Programmer and The Incorruptible Source Code promised with their eyes while they were alive, but they saw things with faith eyes and treated what they saw by faith and believed what they accepted by faith as reality. It was so real that they considered their true homeland out of this world and put all their energy in to getting to this awesome place. This intense Faith Factor made The Powerful, All-Knowing Programmer and The Incorruptible Source Code happy to be their God and call them His people.

Genesis 37:1-11

Abraham's grandson Jacob had an interesting life where he had to work seven years for the girl he was madly in love with. On the wedding night, he got tricked because the wrong girl was behind the veil. So he had to work another seven years for the girl he wanted. Both girls were sisters. The one Jacob was tricked in to marrying gave Jacob ten sons. The wife Jacob loved more gave him two sons. Jacob was closest to these two sons—Joseph and Benjamin. Joseph had a dream thing going on where everyone had to bow down to him. He dreamed it twice. This just made his brothers rage with jealousy. Jacob fussed at Joseph in the beginning but had different thoughts after the second dream.

Genesis 37:12-36

Jacob sent Joseph to check on the brothers who were working out in the fields. They had a major axe to grind with Joseph, beat him up, and threw him in a pit. Then, they sold him to slave traders, killed an animal, covered Joseph's special outfit in blood, lied to Jacob that an animal killed Joseph, and sent Jacob in to mourning. Meanwhile, Joseph got sold to an important dude in Egypt.

Genesis 41:8-16

The top dog in Egypt went by the title of Pharaoh. He tripped out on a dream that no one could explain. Joseph had a knack for doing the right thing even if it got him in trouble which landed him in prison. This knack for doing the right thing was because he had a major Faith Factor going on with The Powerful, All-Knowing Programmer and The Incorruptible Source Code. He practiced the good behavior and morality that Abraham had passed down to Isaac—that Isaac passed down to Jacob—that Jacob passed down to Joseph. While in prison, Joseph's good behavior earned him the respect of everyone including the prisoners and the guards. Joseph earned various responsibilities and even interpreted some dreams correctly. One of the cats he interpreted dreams for remembered his skill even though Joseph gave props to The Powerful, All-Knowing Programmer and The Incorruptible Source Code for giving him the ability to interpret correctly. So the cat with the good memory told Pharaoh about Joseph. Joseph came willingly but insisted The Powerful, All-Knowing Programmer and The Incorruptible Source Code would help him be in the know.

Genesis 41:25-49

The Powerful, All-Knowing Programmer and The Incorruptible Source Code set a plan in motion for seven years of good weather and crops to be followed by seven years of bad weather and starvation. He told this to Pharaoh in coded dreams which The Powerful, All-Knowing Programmer and The Incorruptible Source Code helped Joseph interpret correctly. Joseph not only interpreted correctly, but his major Faith Factor and passion for giving props to The Powerful, All-Knowing Programmer and The Incorruptible Source Code gave him profound wisdom that Pharaoh and all the important peeps caught on to. This resulted in Joseph being given a super important position in the Egyptian government and the plan succeeding so well that there was enough food for the people in Egypt and many other places near Egypt.

Genesis 47:11

The seven years of plenty passed, and everyone in the whole world felt the sting of the famine. Jacob sent Joseph's brothers to purchase grain. They didn't catch that Joseph was their brother because of his Egyptian get-up and aged appearance. Joseph took advantage of this and pried information out of them in the form of interrogations. Eventually, Joseph came clean and got the whole family moved to Egypt where they were able to enjoy some very nice real estate.

3
The Virus Attacks The Lead App

Exodus 1:6-22

Time passed, and there were lots and lots of Abraham's descendants just as The Powerful, All-Knowing Programmer and The Incorruptible Source Code had promised there would be. The new king didn't remember how Joseph saved the day or how Joseph was given the nice real estate for his people to live on. Fearing an uprising, the new king tried to make life difficult on the descendants of Abraham. The king's initial efforts didn't work as planned so the king got all homicidal and wanted the boys killed off. The king's plans still didn't work out perfectly, though, because there were people who had a Faith Factor going on like Joseph did. They chose to do the right thing because of the teachings passed down about The Powerful, All-Knowing Programmer and The Incorruptible Source Code.

Exodus 2:1-15

There was a family that had a boy. They had a major

Faith Factor going on with The Powerful, All-Knowing Programmer and The Incorruptible Source Code. When it became impossible to hide the baby's cries from the king's henchmen, they put the baby in the Nile River in a little basket. The Powerful, All-Knowing Programmer and The Incorruptible Source Code had their back because the king's daughter found the baby, fell in love with the baby, and got the details worked out to raise the baby as her own. To sweeten the deal even more, The Powerful, All-Knowing Programmer and The Incorruptible Source Code worked it out so the baby's real mom could nurse the baby. Little Moses grew up and one day didn't like seeing one of Abraham's descendants being beaten up. Moses went all ninja on the Egyptian and killed him. Moses thought he had covered his tracks, but the word on the street was that Moses was a murderer trying to hide behind his important position as the adopted grandson of the king. Fearing for his life, Moses took off running.

Exodus 3:1-17

Moses got a shepherd gig and a supernatural messenger of The Powerful, All-Knowing Programmer and The Incorruptible Source Code showed up with some news. To authenticate the whole scene, The Powerful, All-Knowing Programmer and The Incorruptible Source Code lit a bush on fire that wouldn't burn up. He gave Moses some important directions about going to Abraham's descendants to tell them He had their back and would deliver them. If they'd follow His lead, He'd help them do some major pest control with all sorts of savage people who had infested the land He had promised to Abraham so long ago. The Powerful, All-Knowing Programmer and The Incorruptible Source Code also gave Moses a name to call Him when Moses went to explain what He had said. The cool thing about the name is that The Powerful, All-Knowing Programmer and The Incorruptible Source Code is always on the scene the way that people who believe in Him need Him to be as I AM.

Exodus 12:12-31

The Powerful, All-Knowing Programmer and The Incorruptible Source Code sent Moses and his brother Aaron to negotiate with the King of Egypt. These negotiations didn't go so well so The Powerful, All-Knowing Programmer and The Incorruptible Source Code sent plague after plague after plague to break the king down. Right before The Powerful, All-Knowing Programmer and The Incorruptible Source Code delivered the death blow on the King of Egypt, He went all detailed and specific giving Moses a game plan for getting out of Egypt and celebrating in a very specific way at the same time every year to remember how The Powerful, All-Knowing Programmer and The Incorruptible Source Code delivered Abraham's descendants from their major struggles in Egypt. (Moses didn't know it, but The Powerful, All-Knowing Programmer and The Incorruptible Source Code were also foreshadowing how the blood of the perfect sacrifice would one day cover people's sin and that spiritual death would not happen if they were covered by the blood!) The death blow involved the first boy born to each family in Egypt dying. This broke Pharaoh down, and he told Moses and Abraham's descendants to leave and worship The Powerful, All-Knowing Programmer and The Incorruptible Source Code as they had asked to do.

Exodus 20:1-17

The Powerful, All-Knowing Programmer and The Incorruptible Source Code got Moses and Abraham's descendants to safety. He summoned Moses up to an important mountain and gave Moses a summary of what He expected out of their behavior. They had to worship The Powerful, All-Knowing Programmer and The Incorruptible Source Code sincerely and not worship anyone else or anything else. If they'd treat name of The Powerful, All-Knowing Programmer and The Incorruptible Source Code with respect plus take the worship rituals seriously, things would go well for them. They also needed to treat the other humans around them the right way by being honest, being responsible, being content with what they have, and not stealing from one another. The Powerful, All-Knowing Programmer and The Incorruptible Source Code took some time to give Moses a bunch of important words, instructions, directions, history, and warnings to write down. Moses took off writing and pretty soon there were five important books—Genesis, Exodus, Leviticus, Deuteronomy, and Numbers. This was literally the point where The Powerful, All-Knowing Programmer and The Incorruptible Source Code began to reveal Himself in writing by inspiring great guys like Moses to write stuff down.

4
The Lead App Recovers
from The Virus Attack

Joshua 6:1-5

Taking back the land that had been promised to Abraham wasn't going to be easy, but The Powerful, All-Knowing Programmer and The Incorruptible Source Code was with Abraham's descendants. A leader named Joshua came on the scene after Moses' time was over. He led the people spiritually and in battle. There were many walled cities in the way. One had a terrible reputation, but if Joshua and Abraham's descendants would follow the directions given by The Powerful, All-Knowing Programmer and The Incorruptible Source Code, they would see the walls of that city fall down. They had to march around the city once a day for six days in a row. Then, they had march around the city seven times on the seventh day. With all the objects of worship The Powerful, All-Knowing Programmer and The Incorruptible Source Code had told them to use in place for proper worship and service, they had victory.

1 Samuel 8:1-7

A lot of time passed. The Powerful, All-Knowing Programmer and The Incorruptible Source Code used a system of judges to govern the people. The people got tired of this and wanted a king. A great spiritual leader named Samuel was upset about this, but The Powerful, All-Knowing Programmer and The Incorruptible Source Code told Samuel not to be so upset about it for the people were not rejecting Samuel, they were actually rejecting the rule of The Powerful, All-Knowing Programmer and The Incorruptible Source Code despite all that He had done for them to help them, love them, and protect them.

1 Samuel 17:1-58

The Powerful, All-Knowing Programmer and The Incorruptible Source Code led Samuel to a man named Saul who became the first king of Israel. Saul had a bad attitude towards The Powerful, All-Knowing Programmer and The Incorruptible Source Code on a regular basis. He disobeyed so much that The Powerful, All-Knowing Programmer and The Incorruptible Source Code took the kingship away from his descendants. The Powerful, All-Knowing Programmer and The Incorruptible Source Code chose to replace Saul with a king named David. David started out as a runt. People even made fun of him. But David had a major Faith Factor going on in The Powerful, All-Knowing Programmer and The Incorruptible Source Code. David's faith was so strong that even when a huge giant made fun of The Powerful, All-Knowing Programmer and The Incorruptible Source Code and Abraham's descendants, David was ready to fight. David couldn't fight the way that normal soldiers do. He fought with faith using the weapons and defenses that had worked for him for many years. Combined with a major leap of faith in The Powerful, All-Knowing Programmer and The Incorruptible Source Code, the giant and his people were defeated.

2 Samuel 5:1-12

After King Saul blew it more than one time by disobeying and disrespecting The Powerful, All-Knowing Programmer and The Incorruptible Source Code, David became the king. David knew Jerusalem was an ideal geographic spot to unite the people of Israel. He was a military genius and knew the walled city could be taken by sneaking in through the water shaft. David and his men went all Navy Seal and overtook the city from the Jebusite people who were there. (The Powerful, All-Knowing Programmer and The Incorruptible Source Code would gradually reveal that He wanted this location to be His center of operations so David's insight and wisdom were really given to him by The Powerful, All-Knowing Programmer and The Incorruptible Source Code.)

2 Samuel 7:1-16

David wanted The Powerful, All-Knowing Programmer and The Incorruptible Source Code to have a temple out of honor and respect. David's major Faith Factor in The Powerful, All-Knowing Programmer and The Incorruptible Source Code was real in his heart, and he wanted worship of The Powerful, All-Knowing Programmer and The Incorruptible Source Code to happen on a grand scale through the implementation of the temple. The Powerful, All-Knowing Programmer and The Incorruptible Source

Code responded by having a prophet tell David that he didn't need a special building because He is so powerful and awesome. However, The Powerful, All-Knowing Programmer and The Incorruptible Source Code also instructed the prophet to say that David's descendants would permanently have the throne because of how David gave props to The Powerful, All-Knowing Programmer and The Incorruptible Source Code.

1 Kings 8:1-13

David's son Solomon had the privilege of building the huge temple as a headquarters of worship and service to The Powerful, All-Knowing Programmer and The Incorruptible Source Code. One important thing that happened in the temple was the sacrificing of animals. Each sacrifice symbolized and pointed ahead to the perfect sacrifice that The Powerful, All-Knowing Programmer and The Incorruptible Source Code would one day accomplish which would cover all the sin, disobedience, mistakes, and evil that people have such a knack for. Israel was all set with much of the land that The Powerful, All-Knowing Programmer and The Incorruptible Source Code had promised to Abraham so many years before plus the temple to help Abraham's descendants focus their worship and thoughts on The Powerful, All-Knowing Programmer and The Incorruptible Source Code. The Powerful, All-Knowing Programmer and The Incorruptible Source Code had called Abraham to be the father of a great nation whose purpose was to carry out the great and wonderful things that The Powerful, All-Knowing Programmer and The Incorruptible Source Code wanted to accomplish in the world.

Sadly, even with the land in place, the temple in place, the worship in place, and Abraham's descendants being very well taken care of by The Powerful, All-Knowing Programmer and The Incorruptible Source Code, they made many mistakes as a people and failed to carry out the plans of The Powerful, All-Knowing Programmer and The Incorruptible Source Code. They showed horrible disrespect and lack of gratitude to The Powerful, All-Knowing Programmer and The Incorruptible Source Code. The Powerful, All-Knowing Programmer and The Incorruptible Source Code sent them prophet after prophet to warn them that they must do a 180 degree turn back towards Him and His ways. They ignored warning after warning and even became a divided nation. Eventually, the rulers of great world empires took them from their land and forced them to live in captivity in foreign places. The sieges were so bad that the land was ruined, Jerusalem was destroyed, and the temple was demolished.

Daniel 7:1-14

Daniel was another cool cat with a major Faith Factor going on in The Powerful, All-Knowing Programmer and The Incorruptible Source Code even in a foreign land where he and Abraham's descendants lived in captivity. He prayed and prayed to The Powerful, All-Knowing Programmer and The Incorruptible Source Code seeking the truth and praying to be able to return to the land that The Powerful, All-Knowing Programmer and The Incorruptible Source Code had promised to Abraham so many years before. Like Abraham's great grandson Joseph, he gained a lot of respect for showing props to The Powerful, All-Knowing Programmer and The Incorruptible Source Code and wound up in important government positions. With the help of The Powerful, All-Knowing Programmer and The Incorruptible Source Code, Daniel had already correctly interpreted the dreams of his king just as Joseph had done for Pharaoh so long ago. Daniel had a dream about four world empires that he saw as beasts that perfectly lined up with the dream that the king had earlier. These dreams symbolized how the Babylonians, the Medo-Persians, the Greeks, and the Romans would all be great world empires and rule over the land promised to Abraham. The Powerful, All-Knowing Programmer and The Incorruptible Source Code would work through these empires to set the stage for the perfect sacrifice to come to fulfillment. Through this perfect sacrifice, The Powerful, All-Knowing Programmer and The Incorruptible Source Code would one day reveal Himself as king and be worshipped. The Powerful, All-Knowing Programmer and The Incorruptible Source Code would also reveal Himself as the great judge of all humanity.

Daniel 9:24-27

Daniel kept praying and praying with all sincerity giving props to The Powerful, All-Knowing Programmer and The Incorruptible Source Code. The Powerful, All-Knowing Programmer and The Incorruptible Source Code gave Daniel a very specific prophecy that pointed all the way to the exact moment when the perfect sacrifice would take place. The Powerful, All-Knowing Programmer and The Incorruptible Source Code gave Daniel a timeline organized in to decades of 7 years instead of decades of 10 years. Each decade of 7 was considered a week of years. There would be a 70 week period of time. The math would crunch out to a 490 year period of time. It would begin when a certain king gave a decree to rebuild the city that was so important to Daniel and Abraham's descendants—Jerusalem. Some time would pass before the 490 year period would begin. Daniel gave props to The Powerful, All-Knowing Programmer and The Incorruptible Source Code and got very precise information that has already all been fulfilled except for a few details that are waiting to happen in our future. The intel was so accurate that it did nail down exactly when the perfect sacrifice would take place and who would be that sacrifice. The 7 year period that still must take place will involve the fourth kingdom from the dreams earlier rising again to power and a great ruler coming out of that kingdom. Despite his greatness, this future ruler will insult and degrade The Powerful, All-Knowing Programmer and The Incorruptible Source Code so badly that it will literally cause all the apocalypse stuff to explode in to action.

Nehemiah 1:1-11, Nehemiah 2:1-9

After Daniel's time, there came another cool cat with a major Faith Factor going on in The Powerful, All-Knowing Programmer and The Incorruptible Source Code. This guy known as Nehemiah also found himself in an important position with the king of his day. Nehemiah heard reports from people who had gone back to Israel and Jerusalem only to see the horrid devastation and ruins that the land was in and that Jerusalem was in. A person could be executed for being sad in the king's presence, but Nehemiah acted sad in front of the king anyway. The king reacted positively by asking what was wrong. Nehemiah poured out his heart about his sadness for the condition of the land of his ancestors and their important city. In a miracle purely orchestrated by The Powerful, All-Knowing Programmer and The Incorruptible Source Code, the king made the decree that The Powerful, All-Knowing Programmer and The Incorruptible Source Code told Daniel would happen one day to rebuild Jerusalem. The king went all out giving Nehemiah full power and access to resources to get the job done. Under Nehemiah's leadership, Jerusalem was restored. The truly sincere worshippers and followers of The Powerful, All-Knowing Programmer and The Incorruptible Source Code wanted to go back to the land promised to Abraham and reinstate their elements of worship. Over time, Abraham's descendants occupied the land promised to Abraham once again.

Despite all the second, third, fourth, and millionth chances The Powerful, All-Knowing Programmer and The Incorruptible Source Code gave His people, they rejected Him and His ways. One day, He took a break from speaking to them in big ways and even hit the pause button on giving men His written revelation. A faithful few always worshipped The Powerful, All-Knowing Programmer and The Incorruptible Source Code sincerely, but Israel as a whole didn't really care about The Powerful, All-Knowing Programmer and The Incorruptible Source Code or give Him props at all. All seemed hopeless, but the king's decree set off the clock ticking for the 490 year period that The Powerful, All-Knowing Programmer and The Incorruptible Source Code had explained to Daniel. The best was yet to come, but The Powerful, All-Knowing Programmer and The Incorruptible Source Code decided to lay low for about 400 years.

5

The Programmer Steps In To The lifePhone

Matthew 1:18-25

Somewhere around the 453rd year of the 490 year period of time revealed to Daniel and launched with Nehemiah, The Powerful, All-Knowing Programmer and The Incorruptible Source Code came to Earth in a form that people could see, hear, and learn from. He started out as a baby and experienced the human experience in every way. To make sure he was uncorrupted, he downloaded Himself to a virgin named Mary. This means no human man was the father of the baby. The Powerful, All-Knowing Programmer and The Incorruptible Source Code knew that Joseph, Mary's fiancé, took the rules and regulations very seriously and would wait until the marriage was complete to have sexual relations with Mary. The Powerful, All-Knowing Programmer and The Incorruptible Source Code also knew this Joseph would protect Mary and the download that came forth as the baby they named Jesus (because The Powerful, All-Knowing Programmer and The Incorruptible Source Code had instructed them to name Him Jesus.)

John 1:1-2, Hebrews 1:1-3

In the very beginning, The Incorruptible Source Code was there in perfect synchronization with The Powerful, All-Knowing Programmer. The Powerful, All-Knowing Programmer and The Incorruptible Source Code expressed Himself in the form of Jesus. Jesus didn't stay a baby. He great up to be the world's greatest teacher ever. Everything about Him perfectly represented The Powerful, All-Knowing Programmer and The Incorruptible Source Code stand for and operates in the perfect ways of the lifeOS.

Jesus did a lot of great stuff while He was on Earth. Most of it was concentrated in to a three year period which began 480 years after the king's decree for Nehemiah to rebuild Jerusalem. In order to appreciate how Jesus did things, we've got to wrap our minds around what His attitude was.

Philippians 2:5-8

Jesus (The Incorruptible Source Code) is and was deity—the ultimate deity actually. He didn't get hung up on His status or position. He took on the form of a servant and expressed Himself in a way that we can see, hear, and learn from as the Incorruptible Source Code. Taking on the restrictions and limitations of the human experience, He helped people and put other people's needs first. He did this so passionately that He ultimately gave up His own life for the lives of everyone else to be the perfect sacrifice that was so badly needed!

John 3:1-21

After the Incorruptible Source Code did enough miracles and taught enough to gain a powerful reputation, a respected religious leader came to Him privately and asked a bunch of questions. The Incorruptible Source Code broke it down and said there's the physical life and the spiritual life. Everyone who has ever been born has experienced the physical life. There is a second birth that has to happen, though, where one becomes a new creature and obtains the spiritual birth. The Powerful, All-Knowing Programmer and The Incorruptible Source Code loved the people of this world so much that He sent Himself in the form of the one of a kind, Incorruptible Source Code. As the son of The Powerful, All-Knowing Programmer and The Incorruptible Source Code, He had the power to be light in the darkness, give people the spiritual birth they needed, and help them have an eternal spiritual life. The spiritual life is readily available for anyone who will sincerely believe and trust in Jesus (The Incorruptible Source Code). Accepting this spiritual life gets the lifeOS back on track in an individual's life. The Powerful, All-Knowing Programmer and The Incorruptible Source Code views the person accepting the spiritual life as completely clean and restored even when the person makes mistakes. The mistakes a person makes causes the lifeOS in the individual's life to go haywire. When a person owns up to the mistakes causing things to go haywire, The Powerful, All-Knowing Programmer and The Incorruptible Source Code goes all tech support and restores the process of the lifeOS working right in that individual's life.

Mark 2:1-12

The Incorruptible Source Code constantly found Himself in situations where people wanted to hear Him teach and wanted to see Him do miracles. One dude was so sick that he couldn't get around at all. His friends were hard corps and lifted him up to the top of the house where the Incorruptible Source Code was teaching. They dug through the first century ceiling and made a hole big enough to lower their friend down to the Incorruptible Source Code. The Incorruptible Source Code was impressed with their Faith Factor and did what He is best at—declaring people free of their guilt and all the entanglements of living life in the messed up ways of this world. Some religious leaders who were supposed to be experts on The Powerful, All-Knowing Programmer and The Incorruptible Source Code were there and cringed about this declaration of "not guilty" that the Incorruptible Source Code made. These religious leaders were supposed to help people know how to give props to The Powerful, All-Knowing Programmer and The Incorruptible Source Code and follow the rules. Instead, they prided themselves on being walking encyclopedias of religious facts and didn't care about the people they were supposed to be teaching and helping. The Incorruptible Source Code saw right through their fake as all get out bad attitudes and decided to prove He can declare "not guilty" by completely healing the sick man physically on top of spiritually. The religious leaders and the Incorruptible Source Code clashed time after time as the Incorruptible Source Code helped people and did the very job the religious leaders were supposed to be doing but weren't doing because they were too hung up on their social statuses, wealth, and power. As time went on, this caused them to plot how to get rid of the

Incorruptible Source Code. They were tech support on paper but not in action. Jesus was Tech Support in action except He can actually kill the virus that's messing up the lifeOS.

John 6:1-15

This is the most popular miracle that the Incorruptible Source Code did. It is recorded four times in the written revelation of The Powerful, All-Knowing Programmer and The Incorruptible Source Code (the Bible) in the sub files of Matthew, Mark, Luke, and John which tell the good news about The Powerful, All-Knowing Programmer and The Incorruptible Source Code wanting to declare people "not guilty" and give them meaning and purpose in life. Andrew was the little brother always living in the shadow of his brother Peter, but he saw the little boy. The Incorruptible Source Code loved to work with the simple and small people—the little guys. So Andrew found the little guy with the loaves and fishes. He told the Incorruptible Source Code's support team about the boy. When the Incorruptible Source Code heard about it, He multiplied the loaves and fishes over and over until nearly 10,000 people had eaten. The written revelation of The Powerful, All-Knowing Programmer and The Incorruptible Source Code says 5,000 men. Since women and children were present, this means a bunch of people enjoyed the feast. They enjoyed it so much that they were ready to make the Incorruptible Source Code their king. The Incorruptible Source Code remembered all the sacrifice business and got away from the crowd like the plague because He had to stay focused on His true mission.

Matthew 18:7-22

As people's misconceptions and wrong approaches to the Incorruptible Source Code grew worse and worse, He intensified His teaching, often teaching things that were difficult for people to want to follow. He told them things like get rid of whatever is hurting your spiritual life and leading you to spiritual destruction. It's better to go without things in the physical life but be prosperous in the spiritual life. The Powerful, All-Knowing Programmer and The Incorruptible Source Code isn't as excited about the majority doing the right thing as He is the minority who have wandered in the wrong direction and need to be rescued. If someone you know does you wrong, go talk to them about it alone and try to work it out. If that doesn't work, get someone else to try to help get it worked out. Still if that doesn't work, you may have to involve the leadership and the sincere worshippers at the worship gatherings. Getting along with each other means forgiving each other over and over—not just one time. This led towards a thinning out of the following, but the popularity was still strong enough to be perceived as a "threat" to the religious leaders who wanted to get the Incorruptible Source Code out of the way because He had stolen their thunder.

Luke 19:28-48

When Daniel prayed and prayed for The Powerful, All-Knowing Programmer and The Incorruptible Source Code to return the people to Israel and Jerusalem, he received the very specific prophecy involving 490 years. There was a specific detail at the 483rd year where the Messiah was recognized as the Prince. This moment when the Incorruptible Source Code is worshipped and praised like a prince is that moment. In the midst of the recognition as the Prince, the Incorruptible Source Code flew into a rage for the second time driving people out of the temple who were just there to make money off the needs of peoples' religious rituals and ceremonies. The popularity as well as the blatant disruption of the religious system of the day, made the religious leaders intensify their plot to get rid of the Incorruptible Source Code.

Mark 15:1-37

The religious leaders who were conspiring and all jealous of the Incorruptible Source Code, manipulated circumstances to the point of the Incorruptible Source Code being arrested and placed on trial. The people who had cheered for the Incorruptible Source Code less than a week earlier had turned completely against Him. They demanded that He be crucified instead of one of the actual criminals. Yet His death was the perfect sacrifice that had to happen to pay for all the damage that had been chain reacting since the first two humans believed the snake and ate the forbidden fruit. While their terrible decision caused The Virus to infect everything, the sacrificial death of the Incorruptible Source Code was the execution of the REPAIR AND RESTORE PROCESS. Anyone who believes and trusts in what He did gets the declaration of "not guilty" plus the NEW IDENTITY WITH REPAIR

AND RESTORE PROCESS enhancing and improving their lives by making them alive for the second birth—for the spiritual birth.

One more cool detail! The 490 year prophecy given to Daniel specifies that the Prince would go through a gruesome death. The Incorruptible Source Code's death on the cross was this gruesome death that happened right on schedule the way it was explained to Daniel several centuries earlier. The "prince" had come to be the king of Israel—the Lead App. However, they rejected Him and killed Him.

John 20:1-9

The Incorruptible Source Code didn't stay dead! Three days later, He came back to life. It took His closest followers a while to fully catch on. The Incorruptible Source Code made various appearances to them until they had seem Him enough to get it that He had come back to life. The way that the Incorruptible Source Code folded the cloth in the empty tomb was the way that a carpenter would fold up his apron and gear when the job was complete. The operation of the perfect sacrifice was complete. The Incorruptible Source Code gave them some final instructions about telling everyone the good news about Him and went back to The Powerful, All-Knowing Programmer and The Incorruptible Source Code in Heaven.

6
The New Lead App

John 14:1-3

Before the Incorruptible Source Code died on the cross, He promised the smaller group of more sincere followers that He was going away to prepare a place for them and would come back to get them. They were familiar with the idea of when a Jewish man got engaged that he would go back to his father's house to prepare a place for he and the bride to live. In the meantime, the bride was supposed to work on the wedding garments. So when the Incorruptible Source Code said this, He was saying that He is the groom and that the believers who remain behind are the bride. The believers are to be preparing their wedding garments for the groom's return.

John 14:15-21, Acts 1:1-Acts 28:31

The Incorruptible Source Code knew that the believers could not truly live the spiritual life without some major spiritual help. So He promised that when He went back to The Powerful, All-Knowing Programmer and The

Incorruptible Source Code that He would send them help—the Spiritual Life Interaction Construct. The Spiritual Life Interaction Construct would help them understand The Powerful, All-Knowing Programmer and The Incorruptible Source Code who is a spiritual being. The Spiritual Life Interaction Construct was actually part of The Powerful, All-Knowing Programmer and The Incorruptible Source Code. He was a form, or expression, of The Powerful, All-Knowing Programmer and The Incorruptible Source Code allocated to help humans function in the spiritual realm with The Powerful, All-Knowing Programmer and The Incorruptible Source Code. One receives the Spiritual Life Interaction Construct at the moment he or she places his or her sincere faith in (Jesus) The Powerful, All-Knowing Programmer and The Incorruptible Source Code. The declaration of "not guilty" plus the NEW IDENTITY WITH REPAIR AND RESTORE process includes the presence and reality of the Spiritual Life Interaction Construct at work within a person.

By the Incorruptible Source Code leaving them in the form they had observed Him in and reuniting completely with The Powerful, All-Knowing Programmer and The Incorruptible Source Code, He could properly initiate the "not guilty" and NEW IDENTITY WITH REPAIR AND RESTORE operations and routines. The allocation of Spiritual Life Interaction Construct would enable The Powerful, All-Knowing Programmer and The Incorruptible Source Code to be inside the hearts of the believers and for them to be in Him. The Spiritual Life Interaction Construct would help them to effectively interact back and forth on a regular basis. This is exactly how the Spiritual Life Interaction Construct works with those sincerely believe and trust in The Powerful, All-Knowing Programmer and The Incorruptible Source Code.

With the Spiritual Life Interaction Construct up and running full force, the number of true believers declared "not guilty" and given the NEW IDENTITY WITH REPAIR AND RESTORE operations and routines expanded tremendously. As the new Lead App, the believers followed the last words of the Incorruptible Source Code and told people how to be declared "not guilty" and run the NEW IDENTITY WITH REPAIR AND RESTORE operations and routines. This developed into organized groups of people called churches. They had to have leaders so the Spiritual Life Interaction Construct gave the new generation of believers the written revelation of instructions about how to do all this. They were to equip the believers to help make other believers and enjoy the maximum benefits of the spiritual life.

Ephesians 2:4-10, Romans 12:1-21, Colossians 3:1-25

These are samplings of the instructions that the Spiritual Life Interaction Construct gave the early church leaders to give the believers. These are all samplings of how The Powerful, All-Knowing Programmer and The Incorruptible Source Code has revealed Himself in writing since the Incorruptible Source Code completed the perfect sacrifice. Believers can practice all these teachings because they have the Spiritual Life Interaction Construct to help them. It is simple stuff like being kind to people, treating other people better than you treat yourself, looking out for children or other vulnerable people in need, forgiving people over and over, keeping anger in control, showing respect for government, showing respect for family members, giving props to parents who care, wearing ways of The Powerful, All-Knowing Programmer and The Incorruptible Source Code like clothes, getting rid of the old and bad stuff, spending time reading and thinking about the written revelation of The Powerful, All-Knowing Programmer and The Incorruptible Source Code, showing sacrificial love, and being thankful. When believers let the Spiritual Life Interaction Construct guide them, these behaviors are a reality. These behaviors also make other people take notice and can lead to chances for other people to hear this good news and get the declaration of "not guilty" plus the NEW IDENTITY WITH REPAIR AND RESTORE process enhancing and improving their lives by making them alive for the second birth—for the spiritual birth.

7

The Complete Restoration of The lifeOS

1 Thessalonians 4:13-17

A spiritual leader in the first 40 to 50 years after The Incorruptible Source Code left Earth to go prepare the place for the Bride was trying to help people who were worried about what happened when people died and when The Incorruptible Source Code would keep His promise to come back to get them. This will happen at a future time when it is the right time in the perfect wisdom of The Powerful, All-Knowing Programmer and The Incorruptible Source Code to return for the Bride. When this occurs, those who died believing will come back to life and head up in the sky along with the believers who are alive when this happens to meet The Incorruptible Source Code in the sky and go to see the place He has prepared for them all.

1 Corinthians 3:8-15

Arriving at the place that has been prepared, the efforts of the believers will be judged to determine the wedding garments that the believers were supposed to be working on until The Incorruptible Source Code came back. Those who were passionate about the ways and the commandments of The Powerful, All-Knowing Programmer and The Incorruptible Source Code will be dressed well and spiritually wealthy. Sadly, those who believed but didn't care so much about doing work for The Powerful, All-Knowing Programmer and The Incorruptible Source Code will be spiritually poor and not dressed as well.

Revelation 4:1-Revelation 18:24, Daniel 9:24-27

While this wonderful seven year marriage celebration is taking place, all sorts of phenomenal, cataclysmic events will take place on the Earth as the ultimate apocalypse. The disappearance of the masses of serious followers of The Powerful, All-Knowing Programmer and The Incorruptible Source Code will result in chaos on the Earth.

A ruler will rise out of the people who destroyed Jerusalem less than 50 years after The Incorruptible Source Code left to go prepare the place for the Bride. Since this was the Roman Empire, this ruler must rise from a revived form of the Roman Empire which could be the European Union or some other form of government that has not yet been formed. This ruler will make Israel, who is always under threat of attack from its neighbors, think they are safe. The covenant he will confirm will make it safe for them to worship and function more freely than how they can right now. They will rebuild the temple and think they are doing the right thing by worshipping The Powerful, All-

Knowing Programmer and The Incorruptible Source Code the old way from before The Incorruptible Source Code came to Earth to teach His ways and die the perfect sacrifice that was required to pay for all the damage caused by The Virus. The risen ruler will turn on them and walk in to the temple that is supposed to be for the worship of The Powerful, All-Knowing Programmer and The Incorruptible Source Code and demand to be worshipped as god instead of The Powerful, All-Knowing Programmer and The Incorruptible Source Code being worshipped as god. This will set off a drastically intensified chain reaction of death, destruction, and all that is bad. Israel's Faith Factor will be pushed to the ultimate testing until they reach the point of believing in the new way to worship that The Powerful, All-Knowing Programmer and The Incorruptible Source Code set up when The Incorruptible Source Code accomplished His mission dying on the cross. Israel will look to the one they rejected and killed with the right attitude just as the 490 years of the prophecy given to Daniel comes to completion.

(The big prophecy given to Daniel was in decades of 7. The 490 years began when the king authorized Nehemiah to go build. 483 years later, the Prince was presented and rejected. The 490 years went on pause with the resume button being the confirmation of false peace with Israel. 490-483 = 7. This is a week [decade of 7]. Jewish weddings lasted a week. So while the last 7 years of the 490 years happens on Earth with the apocalypse, the marriage celebration happens in Heaven.)

Revelation 19:11-21

The Incorruptible Source Code will come back in a very visible form destroying the evil enemies who have been making the Earth a mess and set up a kingdom. This kingdom will honor the promises that The Powerful, All-Knowing Programmer and The Incorruptible Source Code made to Abraham and his descendants. The point of Israel was not to be a physical kingdom so that they could boast of their position with The Powerful, All-Knowing Programmer and The Incorruptible Source Code plus flaunt their wealth and livelihood. The point had been to be a headquarters of good in the world and to lead the way in making the world a better place—to be the Lead App. The Lead App status was transferred to the Bride while she prepared her wedding garments. The position of Lead App opens back up when the Bride (the church) marries the groom (The Incorruptible Source Code). All the plans of The Powerful, All-Knowing Programmer and The Incorruptible Source Code were previously hindered by human shortcomings. With The Incorruptible Source Code as the King, these plans can finally happen on the scale that The Powerful, All-Knowing Programmer and The Incorruptible Source Code wanted them to all along.

Revelation 20:1-15

The Incorruptible Source Code will have the Evil Accusing Deceiver and Corruptor completely bound for a thousand year period while He implements the kingdom and rules it perfectly. The Evil Accusing Deceiver and Corruptor will be released after the thousand year period so The Powerful, All-Knowing Programmer and The

Incorruptible Source Code can prove that some of humanity never wants to worship Him in the proper way. Once released, the Evil Accusing Deceiver and Corruptor will convince these rebels that they can defeat The Powerful, All-Knowing Programmer and The Incorruptible Source Code. This conquest will end terribly for all who oppose The Powerful, All-Knowing Programmer and The Incorruptible Source Code.

2 Peter 3:10-13, Revelation 20:7-Revelation 22:21

This final act of rebellion will be the final violation of the perfection of The Powerful, All-Knowing Programmer and The Incorruptible Source Code. With completely justified rage, The Powerful, All-Knowing Programmer and The Incorruptible Source Code will destroy everything and recreate everything in a pure state—not just a reboot but the entire destruction and recreation of the lifePhone. The complete destruction of the original, corrupted system and recreation will begin and sustain the lifeOS to operate in all the ways that it was originally intended to operate. There will be a New Heaven, a New Earth, and New Jerusalem. The humans of all points in time who had a Faith Factor in The Powerful, All-Knowing Programmer and The Incorruptible Source Code will be alive where they will live, serve, and worship in the huge, awesome kingdom that will last forever. Those who never believed in the teachings of The Powerful, All-Knowing Programmer and The Incorruptible Source Code and who never worshipped Him with the correct Faith Factor will endure eternal conscious torment because they did not allow The Incorruptible Source Code's perfect sacrifice to pay for their acts of wrong and disobedience. Those who get to enjoy the pleasant future are those who were either trusting that this perfect sacrifice would take place to pay for their wrong and disobedience or who lived and believed after the perfect sacrifice took place and trusted that this sacrifice paid for their wrong and disobedience.

8
There's More

If you found yourself connecting with the way life is depicted in The lifePhone, I'd like to invite you to explore what this book is really talking about. This is me using my imagination to summarize and hit the high points of the Bible.

I encourage you to find a Bible and read it. You may want to start with all the Bible passages referenced throughout this book. You can also read the Bible for free at www.biblegateway.com. Most smart phones have free Bible apps.

I also encourage you to find a church to attend and participate in on a regular basis. A church is just a gathering of people who believe in Jesus. It can even happen online, but there is something special about having people to connect with in the physical life.

I've created some online resources to help you go deeper. One is a Facebook page—LifePhone. The other is my You Tube page—WayTruLife Johnny Angel. I will regularly update these with information and ways to follow the teachings of the Bible.

You may also contact me at jangel008@yahoo.com

www.ingramcontent.com/pod-product-compliance
Lightning Source LLC
Chambersburg PA
CBHW071826170526
45167CB00003B/1439